THE BOEING STARLINER MISSION EXPLAINED

Witnessing NASA's First Crewed Boeing Mission and what you need to know about it

Charles D. Battle

Intentionally left blank

The Boeing Starliner Mission Explained

Witnessing NASA's First Crewed Boeing Mission and what you need to know about it.

By

Charles D. Battle

Copyright © 2024 by Charles D. Battle

All rights reserved. No part of this publication may be reproduced, distributed, or transmitted in any form or by any means, including photocopying, recording, or other electronic or mechanical methods, without the prior written permission of the publisher, except in the case of brief quotations embodied in critical reviews and certain other noncommercial uses permitted by copyright law.

This book is a work of nonfiction. The information and opinions expressed in this book are solely those of the author and do not necessarily represent the views or opinions of any individuals or organizations mentioned within.

Gratitude

I am deeply grateful to everyone who has made this book possible.

Firstly, I would like to express my sincere gratitude to NASA and Boeing for their pioneering efforts in space exploration. Their dedication and commitment have been a source of inspiration for this book.

To all the astronauts who have risked their lives for the advancement of space exploration, your courage and determination are beyond words. Your stories have been the backbone of this book.

And finally, to you, dear reader, thank you for your interest in space exploration and for choosing to read this book amongst other books. We have put in so much effort so that this book will satisfy your interest in knowing more about **NASA Boeing Starliner Crew Flight Test** mission. I hope it provides you with a deeper understanding of the Boeing Starliner mission and sparks a sense of wonder about our universe.

Intentionally left blank

Table Of Contents

Introduction...8

Chapter 1: Boeing in the Commercial Crew Race... 14
    Boeing's History in Space Exploration.................... 14
    The Starliner Spacecraft: Design and Capabilities. 16
    The Challenges of Starliner Development (mentioning OFT-1 and OFT-2)............................. 17

Chapter 2: The Crew...22
    Profiles of Astronauts Barry "Butch" Wilmore and Suni Williams... 22
    The Importance of Crew Selection and Experience: Building the Right Stuff for Spaceflight................... 26

Chapter 3: The Mission Breakdown: Launch, Docking, and Return..30
    A Step-by-Step Breakdown of the Starliner Crew Flight Test.. 30
    The Role of the United Launch Alliance Atlas V Rocket...32
    Docking Procedures with the International Space Station (ISS).. 34

Chapter 4: Technology and Innovation: What Makes Starliner Unique.. 36
    Starliner's Advanced Systems: Propulsion, Communication, Life Support.................................. 36
    Starliner vs. Crew Dragon: A Comprehensive Comparison.. 40

Chapter 5: Beyond the Test Flight: Certification and Regular Missions......................................44
    The Significance of a Successful CFT for Boeing and NASA.. 44

    The Path Towards Starliner Certification and
Operational Flights..47
    The Potential Impact on the Space Economy.........51

Chapter 6: Collaborations and International Space Exploration.. 54
    The Role of International Partnerships in the ISS
Program.. 54
    Starliner's Potential for Future International Missions.
56
    The Future of Public-Private Partnerships in Space
Exploration.. 58

Chapter 7: Conclusion: A New Era for Astronaut Transportation..62
    Recap of the Importance of the Starliner Crew Flight
Test.. 62
    The Excitement and Promise of a New Era in Space
Travel... 64
    A Look Towards Future Missions and Discoveries. 66

Bonus Section..68
    The Challenges and Risks of Spaceflight............... 68

Introduction

The Rise of Commercial Crew Programs

In human history, few efforts of some individuals have captured the collective imagination as much as exploring space. From the myths of Icarus yearning to touch the sun to the awe-inspiring images of distant galaxies beamed back by modern telescopes, our fascination with the cosmos has been a constant companion. This yearning to understand the universe and our place within it has driven innovation and scientific discovery for centuries.

The first tentative steps towards spaceflight began with the dreamers and visionaries of science fiction. Jules Verne's "From the Earth to the Moon" and H.G. Wells' "The War of the Worlds" sparked imaginations and fueled scientific curiosity. In the early 20th century, pioneers like Robert Goddard and Konstantin Tsiolkovsky laid the theoretical groundwork for rocket propulsion, paving the way for the practical achievements of the future.

The dawn of the Space Age arrived with a dramatic bang in 1957. The launch of Sputnik 1, the first artificial satellite, by the Soviet Union ushered in a new era of space exploration. The United States responded with the creation of NASA in 1958, marking the beginning of a fierce competition for space supremacy.

The following decades witnessed a series of incredible feats. The first human in space, Yuri Gagarin, orbited Earth in 1961, a moment that captivated the world. The United States responded with Alan Shepard's suborbital flight in 1961 and John Glenn's historic first American orbital mission in 1962. These early missions pushed the boundaries of human endurance and technological capability.

However, the crowning achievement of this era was undoubtedly the Apollo program. This ambitious undertaking aimed to land humans on the Moon, a seemingly impossible dream that became reality on July 20, 1969, with Neil Armstrong. The Apollo missions captured the attentions of the world, it also inspired generations to pursue careers in STEM fields (Science, Technology, Engineering and Mathematics).

Following the Apollo program, NASA's focus shifted towards reusable spacecraft. The Space Shuttle program, launched in 1981, represented a new era in human spaceflight. The Space Shuttle could carry a larger crew and payload into orbit, making it a valuable tool for scientific research and space station construction.

This period also saw the rise of international collaboration in space exploration. The construction of the International Space Station (ISS) began in 1998, a testament to the collective ambition of multiple space agencies, including NASA, Roscosmos (Russia), JAXA (Japan) ESA (Europe), and CSA (Canada). The ISS has served as a crucial platform for scientific research and a symbol of international cooperation towards peaceful space exploration.

The Need for Public-Private Partnerships.

The Space Shuttle program came to a tragic end in 2011 with the loss of the Columbia and Challenger space shuttles. These accidents, coupled with the program's high operating costs and limitations on reusability, forced NASA to reevaluate its approach to human spaceflight.

The retirement of the Space Shuttle left the United States without an independent means of sending astronauts into space. NASA relied on the Russian Soyuz spacecraft for crew transportation to the ISS, a situation that limited mission flexibility and created dependence on another nation.

The need to develop a more cost-effective and sustainable approach to human spaceflight led NASA to explore new avenues. Recognizing the growing capabilities and innovative spirit of the private sector, NASA launched the Commercial Crew Program in 2010.

Introduction of the Commercial Crew Program and its Goals

The Commercial Crew Program (CCP) represents a revolutionary shift in NASA's approach to human spaceflight. It fosters public-private partnerships, leveraging the expertise and resources of both government and industry. This program aims to stimulate innovation and development within the private sector while ensuring the safety and reliability of human spaceflight.

Several leading aerospace companies, including Boeing and SpaceX, competed for CCP contracts to develop safe and reliable crew transportation systems to the ISS. This competition has fostered a spirit of innovation and rapid development within the private space sector. The success of the CCP could pave the way for a future where private companies play a more prominent role in space exploration, while NASA focuses on its core mission of scientific discovery and exploration.

The upcoming launch of the Boeing Starliner Crew Flight Test (CFT) marks a significant milestone in the CCP. This mission will not only be a testament to the capabilities of Boeing's Starliner spacecraft but also a symbol of a new era in human spaceflight, one driven by collaboration, innovation, and a shared vision of pushing the boundaries of human exploration.

Intentionally left blank

Chapter 1: Boeing in the Commercial Crew Race

Boeing's History in Space Exploration

Boeing's involvement in space exploration stretches back to the dawn of the space age, making it a cornerstone of American advancements beyond Earth's atmosphere. The 1960s witnessed the historic Apollo program, a monumental undertaking aimed at landing humans on the Moon. Boeing, alongside industry giants like McDonnell Douglas and North American Aviation, played a crucial role in this endeavor. Their most significant contribution came in the form of the mighty Saturn V rocket, the colossal launch vehicle that propelled astronauts towards lunar orbit.

The Saturn V was a marvel of engineering, boasting five powerful stages, each designed to shed weight and propel the spacecraft further into space. Boeing specifically focused on the first stage, aptly named S-IC, which carried the brunt of the initial launch thrust. This colossal stage, standing 28 meters (90 feet) tall and 10 meters (33 feet) in diameter, housed five F-1 engines – the most powerful rocket engines ever developed at the time. Their combined thrust of 34.7 million newtons (7.7 million pounds-force) was the key to overcoming Earth's gravity and propelling the spacecraft towards the Moon.

Beyond the Saturn V, Boeing's contributions extended to the Apollo program's lunar exploration efforts. They were responsible for developing the Lunar Roving Vehicle (LRV), a lightweight, electric-powered buggy that allowed astronauts to traverse the lunar surface with greater ease. The LRV, used on the Apollo 15, 16, and 17 missions, proved invaluable in expanding the astronauts' exploration range and collecting vital lunar samples. The success of the Apollo program paved the way for the development of the Space Shuttle, a reusable spacecraft designed to ferry astronauts and cargo into low-Earth orbit. Boeing once again played a pivotal role, partnering with Rockwell International in the construction and operation of this groundbreaking vehicle.

The Starliner Spacecraft: Design and Capabilities

The Space Shuttle program revolutionized space exploration by introducing reusability. Unlike expendable rockets, the Space Shuttle could return to Earth, undergo refurbishment, and fly again. Boeing was responsible for designing and constructing the orbiter's aft fuselage and vertical stabilizer, crucial components for flight control and atmospheric re-entry. Additionally, they provided the Shuttle Carrier Aircraft (SCA), a specially modified Boeing 747 that ferried the orbiter back to launch sites after missions.

The space industry continued to evolve throughout the 21st century, with a growing emphasis on public-private partnerships. In 2006, Boeing joined forces with Lockheed Martin to form the United Launch Alliance (ULA). This strategic partnership combined Boeing's expertise in propulsion systems with Lockheed Martin's experience in launch vehicles. ULA has emerged as a leading provider of launch services for both NASA and the U.S. Department of Defense. Their workhorse rockets, Atlas V and Delta IV, have successfully launched a diverse range of payloads, including satellites, probes, and critical cargo resupply missions to the International Space Station (ISS).

Boeing's commitment to space exploration extends beyond launch vehicles. The company continues to develop innovative spacecraft and technologies, with the CST-100 Starliner program being a prime example. This next-generation crew capsule signifies Boeing's continued pursuit of pushing the boundaries of human spaceflight.

The Challenges of Starliner Development (mentioning OFT-1 and OFT-2)

The journey of Boeing's Starliner spacecraft from concept to operational crew capsule has been fraught with technical hurdles and setbacks. This section delves deeper into the challenges encountered during the Starliner's development, focusing specifically on the issues that plagued the Orbital Flight Test (OFT-1) mission and the lessons learned that led to a (mostly) successful OFT-2.

OFT-1

The inaugural Orbital Flight Test (OFT-1) in December 2019 was intended to be a critical demonstration mission, paving the way for crewed Starliner flights. However, the mission was marred by a series of software glitches that prevented the spacecraft from docking with the International Space Station (ISS) as planned. Understanding these glitches and the subsequent corrective actions is crucial in appreciating the complexities of Starliner's development.

1. Navigation Errors: One of the most significant issues encountered during OFT-1 was a malfunction within the spacecraft's Mission Elapsed Timer (MET). This timer plays a vital role in coordinating various critical maneuvers throughout the flight. A software coding error caused the MET to deviate from the planned timeline, throwing off the entire flight sequence. This deviation impacted the spacecraft's ability to perform crucial maneuvers like orbital insertion and rendezvous with the ISS.

2. Attitude Control Issues: Compounding the navigation problems were issues with the Starliner's attitude control system. The software responsible for maintaining the spacecraft's orientation malfunctioned, leading to unexpected thruster firings. This erratic behavior could have jeopardized the spacecraft's stability and posed a potential safety risk for future crewed missions.

3. Data Transfer Woes: Communication between the Starliner and ground control was also disrupted due to software problems. Critical telemetry data, essential for monitoring the spacecraft's health and performance, was either incomplete or delayed. This lack of real-time information significantly hampered the ability of mission controllers to effectively assess the situation and take corrective actions.

The Road to OFT-2

The failures encountered during OFT-1 forced Boeing and NASA to take a significant step back and thoroughly analyze the root causes of the software glitches. A dedicated investigative team was formed, combing through lines of code and scrutinizing every aspect of the spacecraft's software architecture. This intensive effort yielded a series of corrective actions:

1. Software Overhaul: A significant portion of the Starliner's flight software underwent a meticulous review and rewrite process. Engineers focused on identifying and patching the specific bugs that caused the malfunctions during OFT-1. Additionally, the software architecture was re-evaluated to improve redundancy and ensure a more robust system capable of handling potential anomalies.

2. Enhanced Testing: Recognizing the limitations of ground-based simulations, Boeing implemented a more rigorous flight test regime. This involved testing the software in hardware-in-the-loop simulations, replicating real-world flight conditions with actual flight hardware to unearth any potential software-hardware compatibility issues.

3. Communication Revamp: The communication protocols between the Starliner and ground control were also revamped. Redundant communication channels were established to ensure a steady flow of critical telemetry data, even in the event of one channel malfunctioning.

These corrective actions culminated in the launch of Orbital Flight Test 2 (OFT-2) in May 2022. OFT-2, while not entirely without issues, represented a significant step forward for the Starliner program. The spacecraft successfully docked with the ISS, demonstrating its docking capabilities and life support systems.

Intentionally left blank

Chapter 2: The Crew

Profiles of Astronauts Barry "Butch" Wilmore and Suni Williams

The upcoming Starliner Crew Flight Test marks a significant milestone in human spaceflight, not only for showcasing Boeing's Starliner spacecraft but also for the experienced astronauts entrusted with piloting this historic mission.

Barry "Butch" Wilmore:
A seasoned pilot with a passion for exploration
Born in Murfreesboro, Tennessee in 1962, Barry Eugene Wilmore, better known by his callsign "Butch," embodies the spirit of exploration and dedication that defines NASA's astronaut corps. His journey to space began with a distinguished career in the United States Navy, where he honed his skills as a test pilot. Wilmore's expertise in piloting various aircraft proved invaluable, as evidenced by his participation in the development of the T-45 Goshawk jet trainer, a crucial aircraft for training future generations of naval aviators.

Wilmore's exceptional piloting skills and unwavering dedication caught the attention of NASA, leading to his selection as an astronaut candidate in July 2000. The years that followed were marked by rigorous training, preparing him for the challenges and complexities of spaceflight. This training culminated in his first spaceflight in November 2009, a pivotal moment in his career.

As the pilot of Space Shuttle Atlantis for mission STS-129, Wilmore played a critical role in delivering essential supplies and equipment to the International Space Station (ISS). This 11-day mission not only marked his debut in space but also showcased his leadership and teamwork skills as he collaborated seamlessly with his fellow crew members to ensure the mission's success.

Wilmore's dedication to space exploration extended beyond his first mission. He served as a vital member of Expedition 41 to the ISS, further solidifying his experience in living and working in low-Earth orbit. This extended stay on the space station allowed him to contribute significantly to scientific research and participate in various spacewalks, demonstrating his proficiency in performing complex tasks in the harsh environment of space.

Sunita "Suni" Williams:

A record-breaking spacewalker and a scientific mind Sunita Lyn Williams, born in Euclid, Ohio in 1965, is a name synonymous with achievement and perseverance in the world of space exploration. Her unwavering commitment to space science and exploration is a testament to her dedication to pushing the boundaries of human knowledge. Williams' journey began with a distinguished career in the United States Navy, where she served as a naval officer, laying the foundation for her future accomplishments.

Inspired by the vastness of space and the pursuit of scientific discovery, Williams set her sights on becoming an astronaut. Her exceptional academic background, coupled with her experience as a naval officer, made her a compelling candidate for NASA's astronaut program. Williams' first spaceflight arrived in 2006 when she served as a member of Expedition 14 and Expedition 15 on the International Space Station. During this extended stay, she not only contributed significantly to scientific research conducted onboard the ISS but also etched her name in history. Williams became the record holder for most spacewalks conducted by a woman, a remarkable feat that showcased her technical expertise and unwavering dedication to the mission.

Her record-breaking achievement wasn't the only testament to her skills. Williams further solidified her position as a leading spacewalker by amassing over 50 hours of cumulative spacewalk time, a record for women at the time. This experience not only cemented her reputation as a skilled spacewalker but also highlighted her ability to remain calm and collected under pressure, a crucial quality for any astronaut venturing outside the protective confines of the spacecraft.

In 2012, Williams returned to the ISS, this time serving as a flight engineer on Expedition 32. Her leadership skills and vast experience proved invaluable, as she played a critical role in ensuring the smooth operation of the space station. Her dedication and expertise were further recognized when she was chosen to be the commander of Expedition 33, making her the second woman to hold this prestigious position.

The Importance of Crew Selection and Experience: Building the Right Stuff for Spaceflight

The human spirit of exploration thrives on venturing beyond the known. Spaceflight, pushing the boundaries of human existence, places a unique set of demands on those who undertake such journeys. While technological advancements have revolutionized spacecraft capabilities, the human element remains paramount for mission success.

Choosing the right crew for a spaceflight transcends simply selecting technically proficient individuals. Imagine a team of astronauts isolated in a confined space for months, facing the physical and psychological challenges of microgravity and distance from loved ones. The crew selection process becomes a meticulous endeavor designed to identify a well-rounded team with complementary skill sets and personalities.

Here's a closer look at the key pillars considered during crew selection:

Technical Expertise: A strong foundation in scientific and engineering disciplines is essential. Mission requirements determine the specific skillsets sought. Piloting a spacecraft demands expertise in aerospace engineering and piloting experience. Missions focused on scientific research might prioritize astronauts with backgrounds in physics, geology, biology, or medicine.

Physical Fitness: The human body undergoes significant physiological changes in space. Microgravity weakens muscles and bones, while exposure to radiation and a confined environment add further stresses. Meticulous medical evaluations assess an astronaut's cardiovascular health, tolerance for extended periods in microgravity, and overall physical fitness. Additionally, the selection process considers factors like height, weight, and agility to ensure compatibility with the spacecraft's design and potential extravehicular activities (EVAs) – spacewalks – during the mission.

Psychological Resilience: Space travel presents a unique set of psychological challenges. Astronauts confront isolation, confinement, and long periods away from loved ones. In-depth psychological evaluations assess an astronaut's mental stability, ability to cope with stress and boredom, and adaptability in unpredictable situations. Emotional intelligence, the ability to manage emotions and understand those of others, is also crucial for maintaining a positive and productive team environment.

Interpersonal Skills: Effective communication and teamwork are the bedrock of a successful mission. The crew will spend months living and working in close quarters, requiring exceptional collaboration skills. The selection process evaluates an astronaut's ability to communicate clearly, manage conflict resolution effectively, and maintain a positive and productive team environment. Good leadership qualities and the ability to inspire and motivate others are highly sought-after traits.

Intentionally left blank

Chapter 3: The Mission Breakdown: Launch, Docking, and Return

A Step-by-Step Breakdown of the Starliner Crew Flight Test

The Starliner Crew Flight Test (CFT) marks a momentous occasion, not just for NASA's Commercial Crew Program, but for the future of human spaceflight. This chapter delves into the intricate details of the mission, from the launch pad to the return capsule, providing a comprehensive understanding of each crucial stage.

Pre-Launch Preparations: Weeks before liftoff, a flurry of activity takes place at Cape Canaveral Space Force Station. The Atlas V rocket undergoes rigorous checks, ensuring all systems are functioning flawlessly. Meanwhile, the Starliner spacecraft is meticulously inspected and loaded with essential supplies for the astronauts. Wilmore and Williams train intensively in simulators, rehearsing every aspect of the mission, from launch procedures to potential emergencies.

Liftoff and Ascent: On launch day, anticipation hangs heavy in the air. With a plume of fire and a thunderous roar, the mighty Atlas V rocket ignites, propelling the Starliner spacecraft skyward. The initial ascent is the most critical phase, subjecting the crew to intense G-forces as they accelerate towards space. The solid rocket boosters (SRBs) detach after providing a powerful boost, followed by the separation of the first stage booster. The Centaur upper stage ignites, taking over the task of propelling the spacecraft towards its desired orbit.

Orbital Maneuvers and Rendezvous: Once in space, the Starliner spacecraft performs a series of crucial maneuvers. These maneuvers fine-tune its trajectory, ensuring it reaches the correct orbital plane and altitude to rendezvous with the International Space Station (ISS). Throughout this phase, the crew closely monitors the spacecraft's systems and performs pre-programmed thruster burns to adjust its course.

Docking with the ISS: As the Starliner approaches the ISS, a complex rendezvous sequence unfolds. Using sophisticated guidance systems and onboard cameras, the spacecraft meticulously aligns itself with the station's docking port. The crew meticulously monitors critical parameters like distance, closing speed, and relative attitude. Once perfectly positioned, the Starliner extends its docking probe, making contact with the drogue cone on the ISS. After a series of checks to confirm alignment and stability, the spacecraft is carefully maneuvered closer until a hard dock is achieved, creating a secure and airtight connection between the two vessels.

The Role of the United Launch Alliance Atlas V Rocket

The Atlas V, a workhorse of the U.S. launch fleet, serves as the powerful muscle behind the Starliner mission. This highly reliable rocket boasts a modular design, allowing it to be customized for various missions. Here's a deeper look into its anatomy:

Common Core Booster: The heart of the Atlas V is the common core booster, fueled by a potent combination of liquid oxygen and RP-1 (refined petroleum). This powerful first stage provides the initial thrust needed to propel the spacecraft out of Earth's atmosphere.

Solid Rocket Boosters (SRBs): For added thrust during liftoff, the Atlas V can be equipped with up to five SRBs strapped to the side of the common core booster. These solid-fueled rockets provide a significant burst of power during the initial ascent, helping the vehicle overcome Earth's gravity.

Centaur Upper Stage: Once the SRBs and the first stage booster have completed their burn, the Centaur upper stage takes over. Powered by a single or dual RL-10 engine renowned for its efficiency, the Centaur maneuvers the spacecraft into its final desired orbit.

Payload Fairing: Encapsulating the Starliner spacecraft during launch is the payload fairing, a protective shell that safeguards the spacecraft from the harsh aerodynamic stresses encountered during ascent through the atmosphere. Once the vehicle reaches the thin upper atmosphere, the fairing separates, revealing the spacecraft ready to begin its mission.

Docking Procedures with the International Space Station (ISS)

Docking with the ISS is a delicate and intricate ballet requiring precise maneuvers and flawless coordination between the Starliner crew and the station's inhabitants. Here's Below is the steps involved :

Rendezvous Phase: The Starliner spacecraft closes the distance between itself and the ISS using onboard thrusters and sophisticated guidance systems. Throughout this phase, meticulous calculations and precise thruster burns are essential for maintaining a safe trajectory and proper approach angle.

Soft Docking: Once the Starliner reaches a close proximity to the ISS docking port, it establishes a soft dock. This initial contact involves a gentle extension of the spacecraft's docking probe towards a drogue cone on the station. The drogue cone captures the probe, creating a preliminary physical connection.

Load Attenuation and Hard Docking: Following the soft dock, a series of springs and dampers within the docking mechanism

Chapter 4: Technology and Innovation: What Makes Starliner Unique

Starliner's Advanced Systems: Propulsion, Communication, Life Support

Propulsion: The Starliner relies on a two-part propulsion system for its journey. The first leg utilizes the powerful Atlas V launch vehicle, a proven workhorse in the world of space exploration. The Atlas V features a series of rocket stages fueled by cryogenic propellants, a combination of liquid oxygen (LOX) and liquid hydrogen (LH2). These propellants offer exceptional efficiency, generating massive thrust to propel the Starliner out of Earth's atmosphere.

Once in space, Starliner's own Orbital Maneuvering and Attitude Control System (OMACS) takes over. This system consists of a cluster of smaller engines fueled by hypergolic propellants. Hypergolic propellants ignite spontaneously upon contact, eliminating the need for an external ignition source. While hypergolic propellants are less efficient than cryogenic fuels, their simplicity and reliability make them ideal for maneuvering and attitude control in space.

The OMACS system provides Starliner with the thrust needed for orbital insertion, rendezvous and docking with the International Space Station (ISS), and maneuvering during de-orbit and landing. For added safety and mission redundancy, thep OMACS system incorporates multiple engines. In the unlikely event of an engine failure, the remaining thrusters can still safely complete the mission.

Communication: Starliner boasts a robust communication system, ensuring seamless information flow between the crew and Mission Control on Earth. The primary communication channel relies on S-band and Ku-band radio frequencies. S-band provides a reliable, two-way communication link for critical mission data, telemetry (real-time data on spacecraft systems), and voice communication. Ku-band offers a higher bandwidth, ideal for transmitting high-definition video and large data files, such as scientific research data or crew communications with family back on Earth.

Beyond the primary channels, Starliner incorporates a backup communication system for emergency situations. This system utilizes alternative frequencies or even communication satellites to ensure uninterrupted contact with the crew, no matter the circumstances. Additionally, the Starliner capsule is equipped with high-gain antennas that can track and lock onto communication signals, even when the spacecraft is not perfectly aligned with Earth.

Life Support: Starliner's life support system creates and maintains a safe and comfortable environment for the crew throughout their spaceflight. This intricate system consists of several key components:

-*Atmosphere Control System*: This system ensures a breathable atmosphere within the capsule. It utilizes electrolysis to generate oxygen from water vapor, a process powered by the spacecraft's electrical system. Meanwhile, a separate system removes carbon dioxide exhaled by the crew through a combination of filters and chemical reactions. The system also regulates air pressure within the capsule, maintaining a comfortable environment for the crew.

-Thermal Control System: Space presents a harsh thermal environment, with scorching temperatures during launch and ascent followed by the extreme cold of space. The thermal control system ensures a comfortable temperature range for the crew throughout the mission. It utilizes a combination of radiators to expel excess heat during launch and ascent, and heaters to maintain warmth in the cold vacuum of space. This system also circulates air within the capsule to ensure even distribution of temperature.

-Water Management System: Water is essential for human survival, and Starliner's water management system fulfills this critical need. The system starts with a pre-loaded supply of water for the crew's consumption and hygiene needs. Additionally, a water reclamation system filters and recycles used water, maximizing available resources during extended missions. The system can also process moisture from the crew's exhaled breath to create additional water.

-Waste Management System: Starliner incorporates a waste management system to handle waste generated by the crew. This system employs methods like filtration and containment to ensure a sanitary environment within the capsule. The

specific details of the waste management system may not be publicly available due to privacy concerns, but it is designed to be safe, efficient, and odor-controlled.

Starliner vs. Crew Dragon: A Comprehensive Comparison

While both Starliner and Crew Dragon represent significant advancements in commercial spaceflight, key differences set them apart:

1. Propulsion Systems: The Starliner relies on the powerful Atlas V launch vehicle for initial ascent, while Crew Dragon utilizes the SpaceX Falcon 9 rocket. Both rockets offer exceptional performance, but the Falcon 9 uses a different fuel combination (liquid oxygen and RP-1, a form of kerosene) compared to the cryogenic propellants of the Atlas V.

2. Docking Capabilities: Both Starliner and Crew Dragon are designed to dock with the International Space Station (ISS). However, their docking mechanisms and software differ slightly. Starliner employs a docking system called Nodal Point System (NDS), which utilizes a series of cameras and sensors to guide the capsule towards the ISS docking port. Crew Dragon, on the other hand, utilizes a system called Crew Dragon Docking Adapter (CDBA) that features a more automated approach with a softer docking contact.

3. Crew Interface and Automation: The crew interfaces within Starliner and Crew Dragon offer distinct experiences for astronauts. Starliner features a touchscreen-based interface for crew interaction with various spacecraft systems. Crew Dragon offers a similar interface but also incorporates physical buttons and joysticks for added control during critical maneuvers. In terms of automation, Starliner currently relies more on manual crew control for certain aspects of flight, while Crew Dragon incorporates a higher degree of automation, particularly during docking procedures.

4. Launch History and Upcoming Missions: As of March 28, 2024, Crew Dragon boasts a more extensive launch history compared to Starliner. Crew Dragon has completed numerous successful missions to the ISS, transporting astronauts and cargo. Starliner has completed two uncrewed test flights, with its first crewed mission targeted for the near future.

Here's a table summarizing the key differences between Starliner and Crew Dragon:

Features	**Starliner**	**Crew Dragon**
Launch Vehicle	Atlas V	Falcon 9
In-Space Maneuvering	Hypergolic Propellants	Draco Thrusters (using Hydrazine)
Docking System	Nodal Point System (NDS)	Crew Dragon Docking Adapter (CDBA)
Crew Interface	Touchscreen-based with some physical controls	Touchscreen with physical buttons and joysticks
Automation Level	Lower level of automation	Higher level of automation, especially during docking

Launch History	Two successful uncrewed test flights	Numerous successful crewed and uncrewed missionss
Upcoming Mission	First crewed mission targeted for near future	Ongoing missions to the ISS

Chapter 5: Beyond the Test Flight: Certification and Regular Missions

The Significance of a Successful CFT for Boeing and NASA

The upcoming NASA Boeing Starliner Crew Flight Test (CFT) is more than just a launch; it represents a critical juncture in the history of human spaceflight. Scheduled for May 2024, this mission will see veteran astronauts Butch Wilmore and Suni Williams embark on a pivotal journey to the International Space Station (ISS) aboard Boeing's Starliner spacecraft. The success of this mission carries immense significance for both Boeing and NASA, marking the culmination of years of dedicated effort and paving the way for a new era in American space exploration.

A Testament to Persistence and Innovation.
A successful CFT will be the first crewed flight of the Starliner, a milestone signifying the successful translation of years of design, engineering, and

rigorous testing into reality. Following the initial setbacks and delays encountered during the Starliner program's development, a successful CFT will stand as a testament to the unwavering commitment and innovation demonstrated by Boeing and NASA. It will represent the successful resolution of complex technical challenges, showcasing the ingenuity of both organizations in overcoming obstacles and pushing the boundaries of spacecraft design.

Securing America's Path to Low-Earth Orbit.
Beyond its symbolic importance, the CFT holds strategic value for NASA's Commercial Crew Program. This program aims to establish a sustainable and reliable human spaceflight transportation system to the ISS, fostering American independence in low-Earth orbit (LEO) access. Currently, SpaceX's Crew Dragon serves as the primary vehicle for transporting astronauts to the ISS. A successful Starliner CFT will provide NASA with a second American-made, crew-rated spacecraft, ensuring redundancy and mitigating reliance on any single provider. This not only enhances mission flexibility but also strengthens overall national security by providing an alternative

launch option in the event of unforeseen circumstances.

Unlocking the Potential of the Starliner.
The CFT mission serves as a critical stepping stone towards unlocking the full potential of the Starliner spacecraft. Beyond transporting astronauts to the ISS, the Starliner is envisioned for future missions, potentially including free-flying missions in LEO or even crewed lunar flybys. A successful CFT will validate the Starliner's capabilities and pave the way for further development and exploration of its potential applications. This could lead to the expansion of human spaceflight opportunities, fostering a more robust and versatile American presence in space.

Boosting International Collaboration.
The success of the Starliner CFT has the potential to expand international collaboration in space exploration. The International Space Station serves as a unique platform for scientific research and technological development, fostering cooperation between nations around the world. With Starliner as a viable transportation option, more international partners could potentially participate in ISS

missions, enriching the collaborative spirit and accelerating scientific progress aboard the orbiting laboratory.

Inspiring a New Generation.
The upcoming Starliner CFT holds the power to inspire a new generation of scientists, engineers, and space enthusiasts. Witnessing the launch and successful operation of this spacecraft will serve as a powerful reminder of human ingenuity and our collective ability to push the boundaries of exploration. This inspiration has the potential to ignite a passion for science, technology, engineering, and mathematics (STEM) education, fostering a future generation of innovators who will propel humanity further into the cosmos.

The Path Towards Starliner Certification and Operational Flights

The journey from a successful CFT to regular Starliner missions involves a series of rigorous evaluations and certifications. Understanding this process is crucial to appreciating the comprehensive measures taken to ensure crew safety and mission success.

Rigorous Testing and Data Analysis.
Following a successful CFT, NASA will embark on a final round of in-depth analyses before officially certifying the Starliner for routine crew missions. This process will involve a meticulous review of data collected during the CFT, encompassing every aspect of the spacecraft's performance, from launch and docking to re-entry and landing. Engineers will scrutinize telemetry data, video recordings, and sensor readings to identify any potential anomalies or areas for improvement.

Addressing Past Challenges and Ensuring Future Success.
A critical element of the certification process will involve the evaluation of data from a recent parachute test. This test aimed to validate the redesigned "soft link" system, a crucial component that had previously contributed to mission delays. A thorough analysis of this test data will ensure the Starliner's parachute system functions flawlessly,

guaranteeing crew safety during the critical re-entry and landing phases.

Mating the Crew and Service Modules.
Once both the crew module, which houses the astronauts, and the service module, responsible for propulsion and power generation, have undergone rigorous individual testing and certification, they will be mated together to form the complete Starliner spacecraft. This mating process itself will be subject to meticulous inspection to ensure a seamless and secure integration of the two modules.

Setting Clear Milestones and Maintaining Transparency.
NASA and Boeing will establish a well-defined timeline for the certification process, outlining each critical milestone and the associated testing procedures. This transparency will not only foster public confidence in the Starliner program but also ensure clear accountability throughout the certification process. Regular updates on progress and any emerging challenges will be crucial in maintaining transparency and building public trust.
Independent Review Boards and the Power of Collaboration

An essential component of the certification process involves the participation of independent review boards (IRBs). These boards consist of independent aerospace experts who meticulously assess all aspects of the Starliner program, from design and testing procedures to crew training protocols. The IRBs provide an objective and critical perspective, identifying potential risks and offering recommendations for improvement. Their involvement fosters a collaborative environment, leveraging the expertise of various stakeholders to ensure the Starliner's operational readiness.

Addressing Public Scrutiny and Building Trust.
The Starliner program has faced public scrutiny in the past due to development challenges and delays. A successful CFT, coupled with a transparent and rigorous certification process, can go a long way in rebuilding public trust. Open communication regarding the lessons learned from past setbacks and the corrective actions taken will be crucial in this endeavor. By demonstrating a commitment to safety and a culture of continuous improvement, NASA and Boeing can regain public confidence in the Starliner program.

The Potential Impact on the Space Economy

The successful launch and operation of the Starliner spacecraft have the potential to significantly impact the burgeoning space economy. This section delve into the potential ramifications for various stakeholders within this dynamic and rapidly evolving market:

1. A Catalyst for Competition and Innovation.

The Starliner's entry into service will introduce a crucial element of competition into the commercial spaceflight sector. With SpaceX's Crew Dragon already established, the presence of a viable alternative will incentivize both companies to continuously improve their spacecraft and services. This competitive environment can foster innovation, leading to advancements in spacecraft design, operational efficiency, and potentially even cost reductions.

2. Expanding Market Opportunities and Fostering Collaboration.

The presence of two American-made, crew-rated spacecraft can open doors to new market opportunities within the space economy. With increased transportation options, private companies and research institutions may be more inclined to invest in space-based ventures, potentially leading to a surge in scientific research, technological development, and space tourism endeavors. Moreover, the availability of multiple spacecraft could facilitate collaboration between private spaceflight companies and government agencies, leading to the creation of a more robust and multifaceted space ecosystem.

3. The Evolving Role of Private Spaceflight.
The Starliner program represents a significant step forward in the evolving role of private spaceflight. This public-private partnership between NASA and Boeing demonstrates the potential for collaboration between government agencies and private companies to accelerate space exploration endeavors. The success of the Starliner program could pave the way for further public-private partnerships, potentially leading to the development of more advanced spacecraft and the exploration of deeper reaches of space.

4. Uncertainties and Long-Term Considerations.
While the potential benefits of the Starliner program are undeniable, there are also uncertainties to consider. The future of the Starliner beyond its initial NASA missions remains unclear. The long-term viability of the program will depend on its ability to secure additional contracts from NASA or other spacefaring entities. Furthermore, the overall health of the space economy will play a crucial role in determining the long-term success of the Starliner program.

5. A Stepping Stone to a Brighter Future
Despite these uncertainties, the Starliner program signifies a significant leap forward in human spaceflight capabilities. A successful CFT, coupled with a rigorous certification process, has the potential to unlock a new era of space exploration. The Starliner's impact could extend far beyond its role in transporting astronauts, potentially serving as a catalyst for innovation, collaboration, and economic growth within the space economy. As humanity sets its sights further into the cosmos, the Starliner program represents a stepping stone towards a brighter future for space exploration.

Chapter 6: Collaborations and International Space Exploration

The Role of International Partnerships in the ISS Program

THE ISS (International Space Station) stands as a witness to the power of international cooperation in exploring space. Launched in 1998, this orbiting laboratory is a collaborative effort of five space agencies:

1. NASA (United States): NASA brings vast experience in human spaceflight and technological innovation to the table. They contributed major modules like the Unity node and Destiny laboratory, as well as the iconic Canadarm2 robotic arm developed by the Canadian Space Agency.

2. Roscosmos (Russia): Russia's space program boasts a rich history and expertise in spacecraft design and construction. They provided the Zarya and Zvezda modules, which formed the initial core of the ISS, and their Soyuz spacecraft remain the primary vehicle for transporting crews.

3. JAXA (Japan): Japan's contributions include the Kibo laboratory module, known for its robotic arm and science facilities for conducting microgravity experiments in various fields. JAXA astronauts have also played a significant role in conducting research and maintaining the station.

4. ESA (European Space Agency): Representing multiple European nations, ESA's involvement includes the Columbus laboratory module dedicated to life sciences research and the Automated Transfer Vehicle (ATV) that served as a vital cargo resupply spacecraft.

5. CSA (Canadian Space Agency): Canada's contribution focuses heavily on robotics. The Canadarm2 robotic arm is a crucial tool for station assembly, maintenance, and conducting scientific experiments outside the station.

These partnerships have facilitated the success of the ISS in several ways:

1. Shared Resources: Building and maintaining a complex structure like the ISS would be enormously expensive for any single nation. By pooling resources, the costs are spread out, making the project more financially viable.

2. *Expertise Sharing*: Each space agency brings unique expertise and experience to the table. Collaboration allows for knowledge transfer, fostering innovation and problem-solving based on diverse perspectives.

3. *Technological Collaboration*: Joint development of technology allows for faster breakthroughs and the creation of more advanced systems. The ISS itself serves as a testbed for new technologies that benefit future space exploration endeavors.

4. *Global Cooperation*: The ISS stands as a symbol of international cooperation in a challenging and scientifically rewarding field. This collaboration fosters peaceful scientific pursuits and promotes a sense of global unity for a shared future in space.

Starliner's Potential for Future International Missions

The Boeing CST-100 Starliner emerges as a potential game-changer for future international space missions. This next-generation spacecraft boasts features that make it well-suited for collaborative space exploration endeavors:

1. Crew Capacity: With the ability to accommodate up to seven astronauts, the Starliner offers flexibility in crew composition. This allows for international collaboration by enabling a diverse mix of nationalities on a single mission. Scientists and researchers from various countries can participate in space missions, fostering international scientific exchange.

2. Reusability: Unlike previous spacecraft that were expendable, the Starliner is designed to be reusable. This significantly reduces operational costs, making spaceflight more accessible for international partners who may have budget constraints. Reusability ensures a sustainable approach to space exploration, reducing waste and maximizing the use of resources.

3. Docking Capabilities: The Starliner is designed to dock with the ISS, allowing crew rotation and the delivery of critical supplies to the orbiting laboratory. This capability is crucial for maintaining international collaboration on the ISS and potentially future space stations or outposts.

4. Advanced Features: The Starliner incorporates several advanced features to enhance crew safety and comfort. These include a state-of-the-art environmental control system, advanced life support systems, and a modern avionics suite. These features ensure a safe and productive environment for international crews during extended missions.

The Starliner's potential goes beyond the ISS. Its design allows for adaptation to reach various destinations in low Earth orbit (LEO), including potential future commercial space stations or lunar outposts. This opens up exciting possibilities for collaborative space exploration beyond the near-Earth environment.

The Future of Public-Private Partnerships in Space Exploration

Public-private partnerships (PPPs) are poised to revolutionize space exploration. Traditionally, space exploration was dominated by government agencies like NASA. However, the emergence of private space companies like Boeing, SpaceX, and Blue Origin has ushered in a new era of collaboration.

These partnerships offer several advantages:

1. Technological Advancement: Private companies often operate with a more "entrepreneurial" spirit, leading to rapid innovation and development. Collaboration allows NASA and other space agencies to leverage this agility to accelerate technological breakthroughs.

2. Cost Reduction: Private companies typically operate with leaner structures and a focus on efficiency. Partnering with them allows public space agencies to access cost-effective solutions and potentially stretch their budgets further. This is crucial for undertaking ambitious space exploration endeavors that may not be feasible with solely public funding.

3. *Risk Sharing:* PPPs allow for risk sharing between government agencies and private companies. This can incentivize private companies to invest in risky yet potentially high-reward ventures, while mitigating the financial burden for public agencies.

4. *Market Expansion:* Public funding can help private companies develop new technologies and services, fostering a thriving commercial space sector. This can lead to the creation of new markets and economic opportunities, further fueling innovation in the space industry.

Challenges and Considerations for Public-Private Partnerships:
Despite the numerous advantages, PPPs in space exploration also face some challenges:
1. Intellectual Property Rights: Sharing of intellectual property (IP) developed through partnerships can be a point of contention. Clear agreements need to be established regarding ownership and usage rights for new technologies.
2. Mission Objectives: Balancing public and private interests can be complex. Public space agencies may prioritize scientific discovery and exploration, while private companies might focus on profit or commercial ventures. Aligning mission objectives and ensuring transparency are crucial for successful partnerships.
3. Regulation and Oversight: As the private space sector grows, robust regulations and oversight are necessary to ensure safety, ethical conduct, and responsible utilization of space resources.
4. Long-Term Sustainability: Some public-private partnerships might be project-specific. Developing a framework for long-term collaboration and ensuring

a sustainable model for financing future endeavors is crucial.

The future of space exploration is undoubtedly shaped by collaboration. International partnerships and public-private partnerships will continue to play a vital role in driving innovation, expanding access to space, and achieving ambitious goals.
The success of the ISS program serves as a model for future international collaborations. The Starliner, with its potential for international crew participation and reusability, exemplifies how private companies can contribute to global space exploration efforts.
Looking ahead, fostering clear communication, establishing strong legal frameworks, and prioritizing safety and ethical exploration will be essential for ensuring successful collaborations. As public and private entities work together, we can unlock the vast potential of space for scientific discovery, technological advancement, and the benefit of humanity as a whole.

Chapter 7: Conclusion: A New Era for Astronaut Transportation

Recap of the Importance of the Starliner Crew Flight Test

The Starliner Crew Flight Test wasn't just a successful launch and docking; it was a culmination of years of dedicated work by engineers, scientists, astronauts, and countless individuals across Boeing and NASA. This mission held immense significance for several reasons:

1. Validation of the Starliner Spacecraft: The test flight served as a critical validation of the Boeing Starliner spacecraft's design, systems, and capabilities. It demonstrated the Starliner's ability to safely launch astronauts to the ISS, sustain them during the flight, and return them to Earth. This successful test paves the way for future crewed missions to the ISS, providing NASA with a reliable and versatile option for astronaut transportation alongside SpaceX's Crew Dragon capsule.

2. Boosting Public-Private Partnerships: The Starliner Crew Flight Test stands as a potent symbol of the success of public-private partnerships in space exploration. NASA's collaboration with Boeing leveraged the expertise and innovation of the private sector, while NASA provided crucial oversight and mission experience. This model has proven effective in reducing costs and accelerating technological development. The success of the Starliner mission paves the way for future public-private partnerships, fostering further advancement in space exploration endeavors.

3. Reinvigorating Human Spaceflight: The Starliner Crew Flight Test has rekindled global interest in human spaceflight. Witnessing a successful launch and docking after years of development re-energizes the public's imagination and underscores the importance of space exploration. This renewed interest attracts young minds to pursue careers in STEM fields, fueling the next generation of scientists and engineers who will carry the torch of space exploration.

4. Advancing Scientific Research: The Starliner's ability to transport astronauts to the ISS opens doors for a multitude of scientific research opportunities. Astronauts on board can conduct critical experiments in microgravity, furthering our understanding of human physiology, material science, and other disciplines. The Starliner's potential to carry scientists alongside professional astronauts also allows for a broader range of expertise to be brought to bear on research conducted in space.

The Excitement and Promise of a New Era in Space Travel

The successful Starliner Crew Flight Test will mark a pivotal point, ushering in a new era of space travel characterized by several key aspects:

1. Increased Mission Frequency: With both Crew Dragon and Starliner operational, NASA has more options for transporting astronauts to the ISS. This translates to a potential increase in the frequency of crewed missions, allowing for more crew rotation, scientific experiments, and maintenance activities aboard the orbiting laboratory. This increased activity fuels a more dynamic and productive space station environment.

2. Reduced Costs: Public-private partnerships have the potential to drive down the overall costs of space exploration. With the participation of private companies, the financial burden is no longer solely on government agencies. This allows for more efficient resource allocation and opens the door for a wider range of participants, including smaller companies and academic institutions, to contribute to space exploration endeavors.

3. Commercial Space Industry Growth: The success of the Starliner mission further spurs the growth of the commercial space industry. Private companies are now actively involved in astronaut transportation, opening doors for new markets and services. This includes the potential for space tourism, suborbital flights for scientific research, and the development of commercially operated space stations in the future.

4. Technological Advancements: Public-private partnerships foster rapid technological innovation in the space industry. The competition and collaboration between companies stimulate advancements in spacecraft design, propulsion systems, life support technologies, and robotics. This accelerates the pace of development and ensures that space exploration efforts benefit from cutting-edge technologies.

A Look Towards Future Missions and Discoveries

Building upon the success of the Starliner Crew Flight Test, we can confidently look towards an exciting future filled with groundbreaking space missions and discoveries:

1. Man Missions to Mars: The Starliner mission serves as a stepping stone towards more ambitious endeavors like man missions to Mars. The experience gained from operating the Starliner and sustaining a crew in space for extended periods is invaluable for planning and executing future long-duration missions to the red planet. The Starliner's potential reusability also contributes to the economic feasibility of such missions.

2. Exploration of the Moon and Beyond: The Starliner, along with other crewed spacecraft under development, can play a crucial role in future lunar exploration missions. Astronauts could utilize the Starliner to travel to a lunar space station or even participate in crewed landings on the Moon's surface. Beyond the Moon, the Starliner's capabilities could potentially be adapted for missions to asteroids or even further into the solar system.

3. Deep Space Exploration: The success of the Starliner mission contributes to the advancement of technologies and infrastructure necessary for deep space exploration. Discovery new things in space will drag more attention to the outside earth and also it will create awareness to more people on earth.

Bonus Section

The Challenges and Risks of Spaceflight

Space exploration, while captivating and full of potential, presents astronauts with a unique set of challenges and risks. These hazards can be broadly categorized into five main areas:

1. Space Radiation: Invisible and silent, space radiation poses a significant threat to astronaut health. Unlike the protective shield of Earth's atmosphere, astronauts in space are exposed to various types of radiation, including:

- **Galactic Cosmic Rays (GCRs):** High-energy particles originating from outside our solar system, with the potential to damage DNA and increase cancer risk.
- **Solar Particle Events (SPEs):** Bursts of energetic particles from the Sun, capable of causing acute radiation sickness and disrupting spacecraft electronics.

The effects of radiation exposure are cumulative, increasing with the duration of a space mission. Risks include:

i. Increased Cancer Risk: Radiation can damage DNA, potentially leading to various types of cancer later in life.
ii. Central Nervous System (CNS) Effects: High doses of radiation can cause cognitive decline, memory loss, and even cataracts.
iii. Cardiovascular Disease: Radiation exposure can damage blood vessels and increase the risk of heart disease.
iv. Immune System Suppression: Radiation weakens the immune system, making astronauts more susceptible to infections.

Mitigating Strategies:
i. Shielding: Spacecraft are designed with layers of shielding materials like aluminum and polyethylene to absorb or deflect radiation particles.
ii. Mission Planning: Optimizing mission trajectories to minimize exposure to solar flares and planning for shelter during SPE events.
iii. Nutritional Support: Diets rich in antioxidants and radioprotective compounds may help mitigate some radiation damage.
iv. Pharmacological Countermeasures: Drugs under development aim to further protect astronauts from the harmful effects of radiation.

2. Isolation and Confinement: Space travel involves long periods of confinement in relatively small, closed environments. This isolation from Earth and loved ones can have a significant psychological impact on astronauts, leading to:

i. Social and Emotional Issues: Astronauts may experience anxiety, depression, and interpersonal conflicts due to the close quarters and limited social interaction.

ii. Disrupted Sleep Patterns: The unnatural light-dark cycles in space can disrupt sleep patterns, leading to fatigue and decreased cognitive performance.

iii. Sensory Deprivation: The lack of familiar sights, sounds, and smells from Earth can contribute to feelings of monotony and boredom.

Mitigating Strategies:

i. Crew Selection and Training: Selecting astronauts with strong psychological resilience and providing them with extensive psychological training can help them cope with the challenges of isolation.

ii. Simulated Missions: Ground-based simulations can help astronauts experience and practice coping mechanisms for isolation and confinement.
iii. Communication Technology: Maintaining regular video and audio communication with loved ones back on Earth helps astronauts feel connected.
iv. Controlled Environment: Creating a comfortable and stimulating environment within the spacecraft with adjustable lighting, music, and opportunities for exercise is crucial.

3. Distance from Earth: As astronauts venture further into space, the vast distance from Earth creates unique challenges:
i. Communication Delays: Radio signals travel at the speed of light, resulting in communication delays that can be frustrating and impact decision-making during critical situations.
ii. Limited Medical Support: Medical emergencies in deep space require astronauts to rely on onboard medical supplies and their own training. Sophisticated medical procedures are not readily available.
iii. Resupply Limitations: Transporting large quantities of food, water, and other supplies over

vast distances becomes increasingly expensive and time-consuming.

Mitigating Strategies:
i. Advanced Communication Systems: Developing new communication technologies that utilize lasers or other means for faster data transmission is crucial.
ii. Telesurgery and Telemedicine: Research is underway to develop remote surgery and medical consultation capabilities for space missions.
iii. Closed-Loop Life Support Systems: Developing systems that recycle water, air, and waste products within the spacecraft can reduce reliance on resupply from Earth.
iv. In-Situ Resource Utilization (ISRU): Future missions may utilize resources found on celestial bodies like the Moon or Mars for water, oxygen, or even building materials.

4. Gravity Fields: The human body is adapted to Earth's gravity. Astronauts experience different gravitational environments during spaceflight, leading to physiological changes:
i. Microgravity: The lack of gravity in space causes bone and muscle loss, cardiovascular

deconditioning, and fluid redistribution in the body. These effects can lead to balance problems, vision impairment, and weakened bones upon returning to Earth.

ii. Artificial Gravity (Centrifugation): Simulating gravity using rotating modules within spacecraft can potentially mitigate some of the negative effects of microgravity, but this technology is still under development.

iii. Planetary Gravity: The gravity on other celestial bodies like Mars is significantly lower than Earth's. This can lead to similar challenges as microgravity, but also poses a risk of altered biomechanics and potential musculoskeletal problems for astronauts spending extended periods on these planets.

Mitigating Strategies:

i. Exercise Regimens: Astronauts perform rigorous exercise routines using specialized equipment on board space stations and spacecraft to minimize bone and muscle loss.

ii. Artificial Gravity Research: Developing and testing rotating habitats or spacecraft modules that simulate Earth's gravity is an ongoing area of research.

iii. Post-flight Rehabilitation: Astronauts require comprehensive rehabilitation programs upon returning to Earth to regain lost bone density, muscle strength, and cardiovascular function.

5. Hostile/Closed Environments: Spacecraft and celestial bodies like the Moon and Mars present unique environmental challenges:

i. Microbial Contamination: Closed environments can harbor and amplify the growth of bacteria and fungi, potentially posing a health risk to astronauts.

ii. Extreme Temperatures: Spacecraft experience wide temperature fluctuations from scorching sunlight to freezing shade. Astronauts must rely on climate control systems for a habitable environment.

iii. Micrometeoroids and Orbital Debris: Tiny particles of rock and man-made debris orbiting Earth pose a collision risk to spacecraft, requiring protective shielding.

iv. Planetary Environments (Moon and Mars): The Moon's surface is exposed to harsh solar radiation and extreme temperatures. Mars has a thin atmosphere with a low oxygen content and frigid temperatures, making it unsuitable for breathing without protective gear.

Mitigating Strategies:
i. Environmental Control Systems: Spacecraft employ complex air and water filtration systems to maintain a clean and breathable atmosphere.
ii. Shielding and Micrometeoroid Protection: Multi-layered shielding protects spacecraft from micrometeoroids and orbital debris.
iii. Planetary Surface Suits: Astronauts venturing onto the Moon or Mars require specially designed spacesuits to provide protection from the harsh environment.
iv. Planetary Habitation Strategies: Developing long-term habitats on the Moon or Mars will likely involve utilizing local resources for radiation shielding and creating pressurized environments suitable for human habitation.

Space exploration, despite its inherent risks, remains a powerful endeavor driven by human curiosity and the desire to push the boundaries of knowledge. By understanding and mitigating the challenges posed by spaceflight, we can pave the way for safer and more successful missions, ultimately allowing us to venture further and explore the vast reaches of our solar system and beyond.

Thanks for reading…

www.ingramcontent.com/pod-product-compliance
Lightning Source LLC
Chambersburg PA
CBHW070402230526
45471CB00006B/2665